Contents

Style in Product Design

Graham Vickers

The Design Council

Published in the United Kingdom by
The Design Council
28 Haymarket
London SW1Y 4SU

Printed and bound in Great Britain by
Bourne Press Ltd, Bournemouth

British Library Cataloguing in Publication Data
A catalogue record for this book is available from the British Library

ISBN 0 85072 277 2

Front cover: Symphony bath, designed by Queensberry Hunt and photographed by Ivor Innes, Anlaby, North Humberside.

Back cover: diagram of soap dispenser designed by T3 (see pages 67-69).

Acknowledgements

The author and publishers gratefully acknowledge permission to reproduce the following pictures: Olivetti (page 9), Braun (pages 17 and 23); CHILDESIGN (pages 19 and 64); Alessi/Penhallow Marketing Ltd (page 26); Wharmby Associates (page 37); Seymour Powell (pages 38, 43 and 49); Henry Dreyfuss Associates (pages 44, 74 and 76); Nick Holland Design Group (page 45); IDEO Product Development (pages 47, 61 and 72); Shimano (Europa) GmbH (page 55); Tecno (UK) Ltd (page 56); Grey Matter Design Consultants plc (page 60); Smart Design (page 63); DEGW T3 Ltd (pages 68 and 69); Josep Lluscà (page 71); David Mellor Design Ltd (page 73). Pictures not listed are copyright of the Design Council.

The author

Graham Vickers is a freelance journalist and well-known contributor to the design press. His articles appear frequently in a number of publications including *Design, DesignWeek* and *Creative Review*. He is also a contributor to *World Architecture* and *The Architects' Journal.*

Introduction

'Style' is one of those words which tends to mean different things to different people. In the broad context of design it can take on a variety of shades of meaning. When applied to product design in particular it can become even more ambiguous. To the fashion-conscious consumer 'style' may be perceived mainly in products which reflect current trends; to the aesthete 'style' can be synonymous with a kind of detached, timeless elegance; meanwhile for the majority of consumers 'style' is simply the most convenient word for the vague sense of charisma which somehow makes one product more seductive than its competitor.

In all of the above cases 'style' would simply seem to be the subjective word for what is left over when a product has been broken down into its essential components. As a definition, this is rather less than helpful. One of the tasks of this book is therefore to try to arrive at a definition of what style in product design actually means by considering a number of examples. Another is to raise questions about why products which seem to perform similar functions end up looking very different. Does style compromise functionality? What exactly does 'functionality' embrace? Simple ergonomics? Marketing matters? Ease of manufacture? It is such issues that will become clarified by considering a variety of examples.

Commerce

At the centre of any discussion about style in product design are issues of creativity and pragmatism which have wider implications for the design profession as a whole. Designers are not fine artists, but it is surprising how often their work is discussed in terms more appropriate to artistic endeavour. Constantly, terms more relevant to fine art crop up in discussions about design. 'Personal vision' and 'purity of concept' are often mentioned in relation to the design of prosaic household objects which in most cases exist only for the most commercial of motives.

The product designer in particular is obliged to accommodate and incorporate a number of considerations, such as manufacturing and market demands, which have little to do with matters of immediate form and function, but products are frequently assessed and explained almost exclusively in those terms. How the design looks, feels and works is certainly crucial, but a product, unlike a work of art, does not exist as a piece in isolation, and the best product designers are usually those with an appetite for bringing equal creativity to all stages of their design's life, from prototype to final multiple form in the appropriate sales context.

Culture

Just as products cannot exist in isolation from the structure that manufactures and sells them, neither can they flourish in a cultural vacuum. Easy to analyse in retrospect, harder to spot in the present tense, all products bear the signs of their times. Today, in our increasingly visually-oriented society, it is quite common for product designers to make self-conscious references to the past, and sometimes, more unwisely, to the future, catering for simple nostalgia, technological optimism or simply a sense of playfulness in the more sophisticated consumer. But even the least allusive product has to exist in the context of a popular culture, and designers who try to ignore this do so at their peril. Received images of 'futuristic' lines have long been conditioned by cinema imagery, from Fritz Lang to Ridley Scott. Meanwhile graphic design, fashion and music videos all condition the visual responses of the consumers who are exposed to them.

In most discussions about a particular piece of product design, we are inclined to concentrate upon the object itself, taking it out of the context of consumer preferences. By examining its shape, construction and the way it functions, we assume that we shall be able to explain its success, or lack of it. We may choose to refer to the overall effect of physical appeal as the object's 'style', but this does not get us very far. How is that style created? Why is it 'better' than another one? What is being brought into play when we make such judgements?

It is not simply consumer products which must reflect contemporary visual manners. Professional products like medical equipment, navigational aids and all manner of other high-technology products are, in the end, selected from a range of options by human beings. Presented with two functionally identical products costing the same, few people will fail to choose what they perceive to be 'the more attractive'. Which returns us to the teasing question, what *is* 'attractive'? What does the highly emotional process of 'liking' involve? How is it anticipated, stimulated and then satisfied by the product designer? These are questions which any discussion about style must aim to answer.

International style

The term 'International Style' once connoted all kinds of intellectual aspirations centred on the supposed universal validity of the Modern Movement. Stripped of those overtones, the term can today be rather more practically applied to the results of a commercially-driven process which has, to a large extent, homogenized international consumer tastes. That is to say, the things you can buy in the department stores of Tokyo, Los Angeles, New York, London, Paris and across western Europe have a great deal in common when they are not in fact identical. And with that most 'different' of countries, Japan, which addresses Western consumer markets with such spectacular success, it might be assumed that to discuss the nuances of product design style in different countries is irrelevant. To a large extent this is true; and yet certain countries, for various reasons, bring to product design an attitude, a way of working and a cultural perspective which is in itself enlightening. Japan and Italy are obvious examples, and, although this book makes no attempt to deal with product design country by country, national characteristics and national approaches to design must inevitably form part of the discussion.

• • •

In attempting to address the questions raised in this introduction, I have tended to choose products whose design is in some way instructive. Many of the points made could have been demonstrated just as easily by other products created by other designers. This book therefore makes no claim to comprehensiveness. It is not a catalogue of 'the best' – nor is it a history of style in product design, but an analysis of what style means today and how it is used. Most subjects are best illuminated by looking and asking questions, and this is the method chosen here. *Style in Product Design* is an attempt to pin down the elusive and crucial element of style by dipping into a range of examples and influences, and examining the means by which designers, in conceiving and shaping their products, sometimes succeed in creating objects that not only do the job, but also excite and attract.

1 In search of style

In order to reach a clearer definition of style, it is worth taking a look at some specific products which have been widely adjudged to be stylistically successful. None of the following examples is particularly new, and that is intentional – the perspective of time allows a more balanced evaluation of most things, and product design is certainly no exception. Even though an element of subjectivity will creep into any assessment of style, some useful patterns will emerge and the criteria by which products are judged are potentially illuminating. Looking at the products which follow will also help to pinpoint what has made them visually appealing; what made them appealing in the context in which they appeared; and why their success has or has not endured.

The Bell telephone

In 1937 American industrial designer Henry Dreyfuss worked with Bell Laboratories to design the Bell telephone. It became a small American icon: classless, functional, ergonomically sound and quietly stylish. The design remained in production for over 20 years and, until the late 1980s, was frequently referred to as a 'timeless' design. This model was in fact predated by the comparable Jean Heiberg's Siemens phone, popular in Britain and elsewhere. The fact that the Dreyfuss model was

The functional and quietly stylish 1937 Bell telephone was often described as a timeless design, benefiting from a long hiatus in technological development

slightly less monumental perhaps reflects the fact that in America the telephone more quickly became a natural part of daily life.

Why should a telephone look or feel any other way? For a long time it rarely did, but with the advent of digitized exchanges, push-button dialling and many new electronic features, it now seems safe to assume that the timeless tag will gradually disappear. Had telecommunications developed at even a tiny fraction of the rate of computer technology, this Bell telephone would have been lucky to remain in production for even one year. What Dreyfuss achieved may now be seen simply as a highly rational expression of what a non-digital telephone handset had to do. An indisputably good piece of industrial design, its reputation was nonetheless greatly inflated by technological circumstance.

The Braun KM321 Kitchen Machine

Dieter Rams' Braun food mixer model KM321, illustrated on page 23, first appeared in 1957. Although considerably less versatile than food processors designed 30 years later, the KM321 would not look unduly out of place among them. In searching for a sleek, unified look, Rams may be considered to have anticipated the design of modern kitchen machines. On the other hand it is at least a strong possibility that his superb design actually conditioned future solutions, so further ensuring its own 'classic' status. The KM321 was an extraordinary product by any standards, especially those of its contemporaries. German rationalism sought to reduce fussy imagery and minimize visual evidence of fragmentary manufacturing techniques. The Braun KM321 embodies an exceptional visual austerity that typifies the approach.

The Valentine typewriter

In 1969 Ettore Sottsass and Perry King designed a typewriter for Olivetti marketed as the 'Valentine'. By the end of the 1960s the feeling that the established social order was changing had permeated consumerism on many fronts. King and Sottsass exploited this in their design, setting out to prove that even typewriters could reflect the new freedom by being bright and radical rather than drab and functional.

Designed for Olivetti, Sottsass and King's bright and radical Valentine typewriter reflected the new social order of the 1960s

Intended to create a popular consumer market for portable typewriters, the Valentine attracted a great deal of attention when it

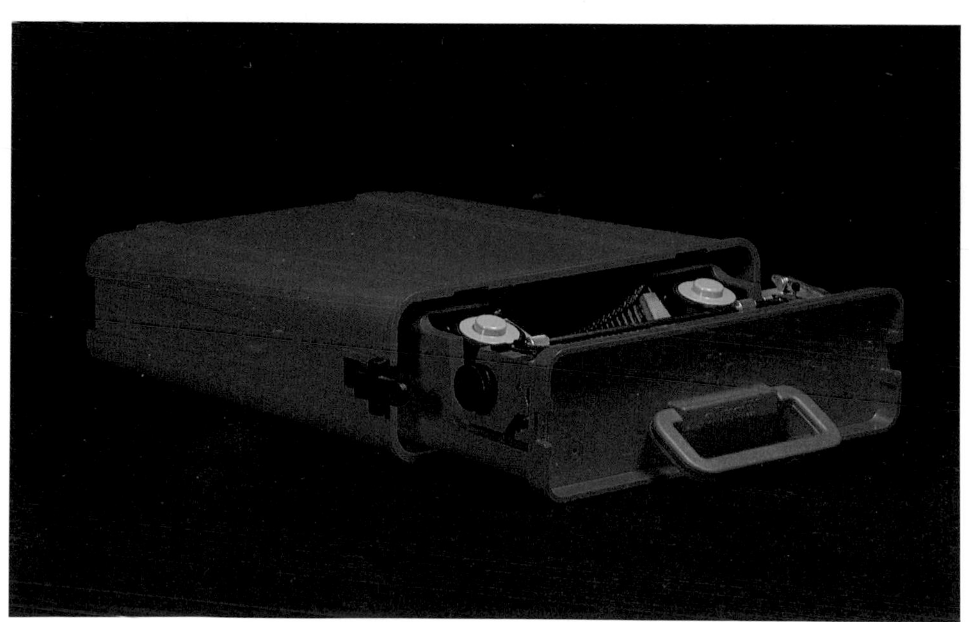

was launched. This was due to its belated appropriation of 'pop' imagery (scarlet finish and matching plastic holster) and its attendant promotion (arresting posters by Push Pin Studios), all paraded under the banner of a prestigious Italian manufacturer. Highly publicized, more or less in tune with its times, stylistically strong but technically weak, the Valentine was a commercial failure.

The Sony Walkman

Launched in 1979, the Sony Walkman has long since passed into legend. Misnamed (in the view of Sony's chairman, who at first thought the name nonsensical) and unwanted (according to the hi-fi dealers), the first Walkman was an adaptation of the Sony Pressman TCM100 tape recorder, launched a year earlier. Despite this rather qualified genesis, Sony had at a stroke created a genuine design original that everyone wanted to buy, instantly and worldwide.

That first model of the Sony Walkman, the TPS-L2, initially marketed as the Freestyle in Australia, the Soundabout in the US and the Stowaway in Britain, established the compact format that countless imitations have since varied only superficially. Later the 'hot button' (suppressing tape signal and amplifying ambient sound) was dropped, as was the second headphone input. The ultimate style accessory for a generation of young people, and now an indispensable travel aid to people of all ages, the Walkman possessed, in the Sony chairman's memorable phrase, 'the potential to fill an as yet unrecognized market need'. At a stroke it reconfigured existing technology into a compact, stylish package that was desirably portable, easy to operate and capable of delivering better quality sound than many domestic systems.

The Alessi Bollitore kettle

Richard Sapper's 1983 Bollitore kettle for Alessi, which, with its added attraction of a whistle mimicking the siren of an American train, will be familiar to many readers, graces many a chic kitchen. It is a product that must have attracted many as an example of stylish 'good' design. But the artful metallurgical confection, beautifully finished and elegantly shaped, initially posed some overheating problems on gas and Aga cookers, although these were later overcome. Perhaps purchasers preferred to sacrifice their own safety for its stylishness.

Style fit for the purpose

The above products have all, at one time or another, been credited with being 'stylish' or displaying 'style'. But the claim may sound more plausible in some cases than others, and no doubt some might be regarded more or less stylish in different periods.

Asked to place the products in order of *design* merit, perhaps many of us would currently place the Walkman first, the food mixer second, the telephone third, the typewriter fourth and the kettle last. And yet for others the kettle would be the most stylish on a purely visual level. Some might disqualify the Walkman as primarily a successful invention rather than a stylish design; others might consider both the telephone and the kitchen machine admirable but hardly imbued with 'style', and the typewriter fun but dated. It could also be argued that the food mixer is the most stylish because its elegant simplicity makes it look timeless, while back in 1969 the Valentine would for some people have come top of a necessarily smaller list.

This range of possible reactions suggests something that many designers resist: that style is not always some magically indivisible part of a piece of good design. It can be little more than the added visual and/or tactile quality which enhances the psychological appeal of a product and which is determined by preferences, taste and changing fashion. As already noted, however, there will always be some products that profit from unexpected longevity, just as there will always be others that date surprisingly quickly.

This does not mean that style is purely a matter of luck. Preferences, taste and cultural influences can be analysed and exploited by designers, and style remains remarkably seductive, a crucial part of a product's viability. But it is unwise for designers or critics to claim too much for style: however much we might like it to be an inextricable part of the product, it rarely is. Of the above examples, only the food mixer can claim a genuine style that is absolutely intrinsic to and indivisible from its form. Its unified look *is* its style.

The Braun food mixer could be judged less successful than the Walkman simply because for many contemporary consumers its style was too intellectual for its time. Braun products rarely sold well in America and usually found a warmer welcome in US museums than in US homes. This demonstrates how the style of a product may be inappropriate for the job in hand. The style of the Walkman, however, is intrinsic to its fulfilment of the consumer's desires and has therefore achieved worldwide commercial success, while the Braun food mixer only ever satisfied a relatively small, style-conscious elite.

2 Cultural context

As consumers, our visual expectations about design are conditioned by all sorts of things. Fashion, television, newspapers, magazines, movies, computer graphics, telecommunications, transport, architecture, advertising . . . these and a wealth of other influences colour our perception of the things with which we choose to surround ourselves, and cause us to 'like' or 'dislike' products such as those discussed in the previous chapter. It follows that consumer acceptance of a particular commodity – a chair, a table, a lamp or whatever – will to some extent be dependent upon how well it accommodates visual and conceptual prejudices already imprinted on consumers by all sorts of other imagery. For example, a young adult male faced with the task of choosing a chair to furnish his room may already have his range of options limited by a range of preferences, such as his taste in cars and computers, magazines and films. The chair does not have to 'go' with any of these things in any interior design sense, but if his predilection is for sleek, black, high-performance cars, professional-looking audio components, high-tech adventure films and fashionable style magazines, he is extremely unlikely to opt for a chintzy look. When he narrows his choice down to three more or less suitable contenders, he will eventually choose the one whose designer has best read his tastes.

That is an obvious example, but the same forces are at work at all levels. If designers want to address consumers in an informed way, it will clearly be necessary for them to take on board a very wide set of popular references. Of course, consumers buy things for all sorts of reasons, and products and commodities will have to succeed on practical levels too – function, price, ease of use – but these may not actually determine commercial success or failure. Faced with a choice of products equally priced and offering similar performance, consumers invariably buy the one that they 'like' best. Liking, if not exactly 'liking what we know', is at least usually a process of being drawn towards something that does not openly challenge existing preferences.

Powers of association

The most casual consumer is affected by cultural associations. If this phrase sounds vague, its meaning can be illustrated by considering the 1982 Sony TC-5550-2 reel-to-reel tape recorder. A domestic product, the TC-5550-2 employed 'professional' imagery (mimicking a Swiss tape recorder used in broadcasting) to enhance consumer appeal. Many people who bought that Sony product would not necessarily have been able to identify exactly why it impressed them, but the clearly received impression that it looked somehow 'serious' is one example of cultural association at work.

The introduction of graphic equalizer controls to portable radio cassette players is another instance of a similar principle. Since such products often have only very basic sound quality, there was no great practical need to replace the single tone control with an array of up to ten sliding controls. However, the popular image of a recording studio mixing console presented a convenient model of technical superiority to be exploited.

The process of association can work much more subtly than in these examples. In less obvious cases to look at an individual piece of design in isolation is of even less value. The broad context in which any product exists is important, and not only to designers, critics and students of design, but also – perhaps most importantly – to the people who hope to make money out of commissioning and manufacturing well-designed products.

Originality and classics

In fact, the question of cultural context is perhaps most sorely ignored not by designers but by the people who employ them. Unimaginative entrepreneurs and industrialists are not just disinclined to take creative risks; if they are out of touch with everything but the balance sheet, they are actually *unable* to take creative risks. Even a superficial industry comparison between the United Kingdom and Italy, or the United States and Japan, reveals that it is not the domestic design talent which varies so dramatically, but the underlying attitudes of the national industries that are deploying it.

A simple analogy is provided by mainstream American cinema. Every now and again a film of some originality becomes a big commercial success. At this point the decision-makers at the major studios (accountants at heart) inevitably attempt to repeat that success by duplicating what they imagine brought it about in the first place – that is, they set about copying it, adding a numeral to the original's title. What is usually copied is a simple pattern of characteristics which might be called the 'style' of the original – but as soon as it is copied, style becomes mannerism, and, no matter how cleverly done, the copy rarely duplicates the original's success. Reducing any original to a simple stylistic formula which can then be repeated is to miss the point. And yet many product designs are created in this way, vaguely seeking to duplicate someone else's success by adapting the imagery.

Everything succeeds or fails in its own right and in its own cultural context – that context is in a permanent state of flux. Indeed, today popular culture changes more rapidly than it has ever done before, modifying and re-aligning our visual expectations at an unprecedented rate. Proof of this can be found in the ephemerality of design styles in many contemporary product areas, notably automobiles, hi-fi and TV sets; even people who are not particularly interested in hi-fi can now usually sense that a two-year-old system – perhaps with components virtually identical to those in the latest model – looks out of time.

It follows that designers who are seduced by or encouraged to copy the style of others (or even of their own previous designs), rather than directly addressing the needs, tastes and preoccupations of the market, are likely to fail at the most elementary levels: firstly, they are distancing themselves from the end user by accepting someone else's rationale; secondly, the product which apes fashionable mannerism may anyway be out of date by the time it appears on the market.

Examples of design longevity can always be found, but whilst certain product designs may eventually become widely considered as classics, they are nearly always the rare offspring of a number of coinciding factors. One of those factors is a certain sort of good design, but the rest are both less controllable and less predictable. When Dieter Rams produced the 1956 Phonosuper SK4 record player for

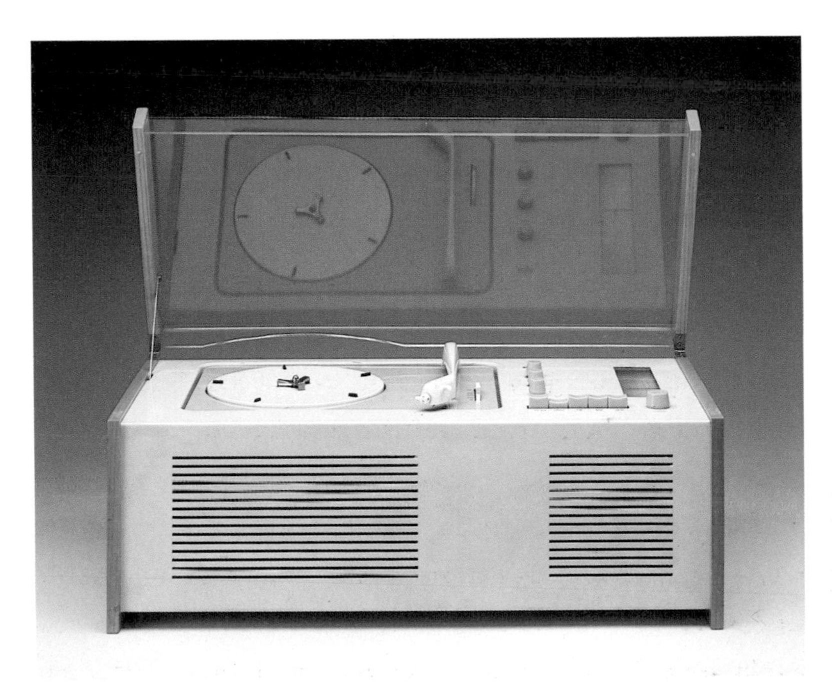

The cool, austere lines of the Braun Phonosuper SK4 record player might have made it a classic but new technology brought about its obsolescence

Braun with Hans Gugelot, he probably imagined its 'classic' simplicity would make it wear as well as his shavers and food mixers. Today the SK4 is visually and practically redundant, and long out of production – even though the Vitsæ shelving system, optimistically built for it, is still available.

On the other hand, just because a designer sets out to satisfy a very pragmatic brief it does not mean that the outcome will not be an enduring product: 'classic' design is simply an irrelevant objective in itself, since the designer has no control over the future context in which the design will be viewed. By contrast, designers are – or should be – extremely aware of the immediate context into which the final product will be launched, although that context may not always be to the personal taste of the designer.

Sometimes the task may involve embracing a nostalgic aesthetic which the client has identified as being commercially successful. The British Marks & Spencer chain of shops is one of the most successful retail organizations in the world. In becoming so, it adopted a very conservative view of design, both in its products and its stores. It nevertheless uses designers, and the design task of developing what may to some people seem bland, middle-of-the-road products is no less difficult than creating radical chairs and kettles which may get no further than the pages of lifestyle magazines. Only if the product designer is seen as a kind of artist-in-waiting can the design of, for example, pseudo-traditional kitchenware be considered intrinsically inferior to that of the boldly innovative product. Both exercises call for the expert manipulation of visual imagery to suit the appropriate commercial and cultural context.

Taste and needs

Looking at designs in their cultural context can be both confusing and helpful. It may be confusing because interpreting cultural signs is often approached in a highly subjective way and such an approach will never in itself endorse a specific solution conclusively. At best it will provide the designer with a relevant palette of visual options and permit manufacturers or entrepreneurs to make better-informed decisions about exactly how they want a design to appeal.

Much more helpful is an approach which encourages the designer to have a more direct relationship with the consumer. Failure to do this may bring about major commercial disasters. For example, when inventor Clive Sinclair launched his disastrous C5 motorized tricycle on the British public in the mid-1980s its logic was unassailable: it was small, cheap, ecologically sound and the perfect hybrid of car and bicycle. However, it was not the C5's subsequent manufacturing problems alone which made it such a disaster. It was Sinclair's isolation from public taste. The image and associations of the C5 were largely comic – for a society addicted to the car and recently seduced by the technological sophistication of video recorders and personal computers, the pod-like C5 simply looked like a cranky throwback to the bubble car of the 1950s. Those seeking the elegance and practicality of a genuinely green machine rode bicycles. Like so many inventors, Sinclair failed to focus on the psychological perceptions of the consumer.

A perceptive eye can always open up design opportunities. It may be that the designer judges the moment right to introduce a note of visual contrast. The Sharp radio cassette player illustrated is proof of this. It represented a conscious bid to challenge the 'black/grey box' aesthetic that at the time ruled the competition. In other situations it may be that a traditional look is required to establish an unfamiliar product in its desired niche. Either way, a clear reading of contemporary imagery is crucial.

A conscious bid to challenge the black/grey box aesthetic: the Sharp radio cassette player

Users' special needs, such as physical disabilities, will also influence the style of a product. Whether it simply works as a piece of basic equipment or manages to be attractive at the same level as other non-specialized products, depends on the visual receptivity of the designer and the manufacturer.

These are all practical issues directly related to the consumer society in which we live. It is out of just such pragmatic requirements that many products of style and distinction have been created. They are, however, greatly outnumbered by products which are manifestly vulgar, impractical, dangerous, offensive or simply foolish. More insidiously, they are matched by products which claim 'style' by pretending to be art objects – expensive, exaggerated and often quite sophisticated bids to play tricks with our expectations and elevate product design into a branch of fine art.

It is the search for that first option, the creation of something seductive and exciting that still performs its intended job, which provides the most interesting insights. These days it can rarely be achieved by designers who are disdainful of popular culture.

3 National style

Everyone's cultural preferences are to some extent determined by national identity. In an increasingly homogenized world it is often the trivial details of an alien environment which catch the attention of the traveller. When you can pick up the same satellite TV news station in London, New York and Tokyo, and buy the same products at airports the world over, small points of distinction can take on sharp relief. Why does American scaffolding use tubing with flattened ends to make the connections whereas the British equivalent uses clamps? Why is the besom considered to be a practical form of yard brush in many European countries when in Britain it is viewed almost exclusively as an antique item? Why do Parisian underground train doors have to be unlatched by the passenger when no such need is perceived in New York, London or Milan? Trivial as these questions are, they touch upon an important principle. There is no apparent reason for the above discrepancies other than that of habit and simple preference. They are the survivors of variety which was once total but which these days is much reduced by modern communications.

When it comes to design and style there are still often deeply rooted traditions of national individuality colouring the way things are perceived and fashioned. At the same time, helping to negate those differences, is an increasingly pan-global marketplace. Somewhere in between, decisions are taken which may seek to reinforce or erase national design distinctions. If we accept that a portable stereo, an electric razor or a calculator has to perform essentially the same functions the world over, it is in the domain of visual style that national differences of emphasis are likely to occur.

It is not the purpose of this book to develop a detailed discussion of national styles in design, but it is worth sketching in a few broad national characteristics which have made significant impact upon design decisions and consumer choice.

Germany

The received image of German product design is in effect a legacy of three establishments:

■ *The Deutsche Werkbund* was an organization formed at the beginning of the twentieth century with the aim of drawing together industry and art. It coincided with a mood of standardization and mass production and drew some affinities from contemporary English design ethics.

■ *The Bauhaus (1919-33),* was an extraordinarily complex and multi-faceted educational establishment with its roots in Expressionism and Arts and Crafts. Despite this, it eventually came to emblemize the impact of the machine age upon art and design, and it is in this symbolic role that its name is most usually evoked.

■ *The Ulm Hochschule für Gestaltung* was a natural successor to both the above. Founded in the years after the Second World War, its agenda was to relate mechanized forms to human requirements – a kind of psychological ergonomics. Even so, its successes are perhaps

The Braun KM321 Kitchen Machine, designed by Dieter Rams, was a product of the Hochschule für Gestaltung and exemplified German functionalism

most easily seen from a contemporaneous German point of view. Today products that emanated from it or which were influenced by it tend to look highly rational and even rather detached in their appeal.

The recurring strand of contradiction in the above brief definitions, is a trait which runs throughout German design and something which lies at the very heart of any discussion about style. It has been suggested earlier that there is no real foundation for the assumption that 'style' is anything but an additive, no matter how cleverly integrated. Many will dispute this and some might hold up German design as a broad exemplar of how bold functionalism can result in potent visual appeal that is indivisible from form. A more accurate assessment may be that whilst German products have frequently combined high quality materials with superb engineering and rational thought, the form that they take is often shrewdly calculated to *project* just these qualities – that is to say it is essentially a styling exercise, albeit a highly subtle one. In general, German products have been no less susceptible to change for change's sake and to embracing a style which is actually irrelevant to function than products from other countries.

The United States

America's most lasting contribution to the language of style remains the flamboyant excesses of 'styling' – a theatrical visual ploy at its height in the 1930s which was intended to stimulate demand for products such as cameras, vacuum cleaners and washing machines. Later, its influence was seen in the absurd (and lethal) tail fins of Cadillacs and other American dream machines (see page 30). As with many American phenomena, there was arguably a kind of virtue in styling, in this particular sense of the word. It was obvious and overstated, and yet it was somehow appropriate imagery for a vigorous nation with boundless enthusiasm to signal its appetites and its love of commerce. The most prosaic products were made to look as if they were capable of travelling to Mars, even toasters, such as the 1950 Westinghouse model, which typified the overblown message of styling.

The enduring legacy of styling is that it has become a kind of visual shorthand for evoking a certain aspect of mythical America; fins, grilles, chrome curlicues and bulbous streamlining have taken their place alongside Levi's, Coca-Cola, the Stars and Stripes, Greyhound buses and other solidly exportable national icons. In *From Bauhaus To Our House* Tom Wolfe has berated America for subsequently being seduced by the grave European prophets of Modernism when what America really needed was a visual vocabulary of exuberance. The best of today's American product design owes nothing to the excesses of styling and rarely possesses any indentifiably 'American' qualities.

Italy

Italy's reputation as being the natural focus of the international design community is unchallenged. Explaining quite why this should have come about is rather more problematic. It seems safe to assume that a uniquely rich cultural heritage plays a highly significant part. So too did the catastrophic decline of Italy's infrastructure, exposed after the end of the Second World War. The ensuing reconstruction of Italy found a place for 'design' as an adjunct to architecture and by the 1950s the designer had become a leading player in many industrial processes, with a role that included distinguishing between rational form-follows-function design and more playful 'marketing' design. The advent of technopolymer materials intrigued many Italian designers who quickly gave them a legitimacy that had nothing to do with plastic imitating other materials; for this development alone Italy can claim to have created a whole new burgeoning aesthetic with immeasurable impact on the style of a whole range of products.

In the 1960s, with Italian design enjoying an international reputation as the natural expression of a materialistic good life, many Italian designers still became dissatisfied. A gradual decline in the

The 1983 Alessi Bollitore kettle by Aldo Rossi; thanks to such products Italy is seen as the unassailable international capital of design

prestige of the designer and a reaction against Italian design's increasingly bourgeois image prompted two renegade 'anti-design' movements: Radical Design and the famous Memphis. These added a spirit of creative devilment to Italy's already heady reputation as the international capital of design.

In terms of national style in product design, 'Italian Design' is now almost entirely meaningless except that daring, playful or outrageous product design is often described as having an Italian look, even though it may come from Japan, Spain or even Britain. One conclusion to be drawn from this is that, as far as Europe goes, it was Italy where cultural history and historical circumstance first combined to demonstrate that product design could be anarchic, outrageous, elegant, surprising, beautiful – or indeed anything. It was a national achievement that would subsequently be challenged only by a nation with a very different history, but one that nevertheless reveals some interesting parallels.

Japan

Like Italy, Japan had to reconstruct itself after the Second World War. However, Japan's spectacular re invention of itself is so total and all-embracing as to have changed the world order in less than 50 years.

Japan had been obliged to open up to the West at the end of the nineteenth century and had therefore already begun a process of reassessing its traditional culture in the light of the demands of the modern world. It is this process of redeploying past values to new ends which has made modern Japanese design peculiarly successful – and occasionally just plain peculiar.

What Japan shares with Italy is an intuitive acceptance of aestheticism as a natural component of civilized life. Unlike Britain, where art is popularly perceived as difficult or elitist and design is still sometimes regarded with suspicion by industry, there have been no such divisions in Japan. This, combined with an intensity of national purpose, has resulted in Japan becoming a potent force in international product design, particularly in electronic goods.

However, there remain conflicting elements in Japanese design, although the resolution of problems continues apace. A fundamental aspect of traditional Japanese design is a kind of elegant austerity which is not adequately translated as 'simplicity'. Embodying a spiritual attitude towards desirable minimalism in all things, it was not a quality which at first seemed to lend itself to foreign markets. One result was that some Japanese design strove to hijack all kinds of elaborate foreign imagery in a bid to demonstrate its willingness to accommodate the unfamiliar.

Increasing confidence and shrewd marketing brought more successful products, with Sony as an early standard-bearer of just what could be achieved: highly expressive products which were desirable and reliable, and designed to be successful worldwide. One of the more fundamental aspects of Japan's contribution to style in product design has been its clear distinction between the various processes of industrial manufacture and design. However, although 'engineering designers' and 'cosmetic designers' may work independently in the very early stages of a product's development, their individual efforts are eventually unified not by some mystical creative process but by the firm management and team spirit that has come to characterize Japanese industry. The pragmatic Japanese approach reinforces the point that style is not some elusive and inseparable part of the designed product, but an ingredient to be carefully deployed and artfully integrated.

Spain

In the 1980s Spain re-emerged as an important force in design. Perhaps the most remarkable aspect of this apparent renaissance was that it was not a renaissance at all. Denied any real commercial contact with the outside world during the Franco years, Spain was obliged to design and manufacture for its own internal market. However, such was the richness and diversity of Spain's visual culture that a process of almost subterranean cultural continuity persisted, centred on educational establishments and assisted by Spanish designers and architects who studied and travelled abroad. The result was that with the restoration of democracy a visual culture reappeared that, contrary to what one might have expected after Franco, had more in common with Italy than the Soviet Union; in some parts of Spain designers can now receive the kind of media treatment more usually associated with actors and pop stars. Spain is consequently poised to be among the most interesting and innovative of design nations as the twenty-first century approaches.

Scandinavia

It is debatable whether Scandinavia can be said to have had a major influence on style in the sense in which we are discussing it. Perhaps its most influential contribution has been to show that traditional crafts could be reinterpreted for modern tastes without necessarily evoking nostalgia. Scandinavia has also encouraged manufacturers – as well as private businesses and public authorities – to make greater use of artists. In Sweden a powerful social ethic demanded that beautiful objects should be democratically available. Even with craft-based industries this was feasible, given the size of the population. Similarly, the Danish furniture industry was able to maintain high craft standards by addressing a relatively small but affluent international market. Seen in the broader perspective of huge global modern markets, these and similar Scandinavian design achievements may look modest, but the humanistic message they were conveying as early as the New York World Fair of 1939 remains potent today.

Blurring boundaries

Although these national tendencies still determine consumer preferences in different countries to some extent, notions of 'national' design, at least in terms of definable style, are getting more confused all the time. Designing objects specifically to appeal to foreign markets is no longer widespread or feasible. No doubt there will always be local habits, practices and conditions which require product modification, but these are not always entirely predictable: personal stereos for the US market are more likely to include radios because of the strength and number of local radio stations there; less logical is the fact that the waterproof beach models of the Sony Walkman sell well not only in California, but also in Canada and Sweden. However, the dominant move is not towards distinction but unification and it seems likely that in future, as differences between national cultures become less distinct, there will be little one can point to as typical of one country or another apart from traditional products which embody their own national stereotypes. It is precisely this blurring of differences that reinforces the popularity of Burberry and Aquascutum clothing as exportable emblems of an aristocratic British way of life, and the continuing iconic appeal of blue jeans or Harley Davidson motorcycles as icons of America. None of these has much to do with contemporary life in either country.

Notions of supposedly 'national' imagery will nevertheless persist and continue to be deployed, though not in the way they once were. Notions of German rationalism and American exuberance show no signs of weakening, even when the recent histories of those nations no longer support the traditional view. 'Scandinavian' and 'Italian' are still thought of as useful adjectives when describing the impression given by certain designed objects, even if they were made in Korea.

A Dutch-based organization, CITE, is engaged in aligning the new technology aspects of design education into courses which are then started simultaneously in several European countries. The Tokyo Design Network – comprising Nissan, Sony, NEC, Canon and Hitachi – was formed in 1991 'for the promotion of design-oriented activities to benefit society' according to its president Hiroshi Shinohara (John Stoddard, 'Tokyo Networking Reveals Its Targets', *Design Week*, 1 November 1991). He adds: 'Product design needs new values in the emerging post-industrial, environmentally-conscious era'. Whatever that gnomic statement actually means, it seems safe to assume that it embraces an increasingly global view of product design.

American exuberance expressed in the form of streamlining: General Motors' 1954 Buick Le Sabre, designed by Harley Earl (shown at the wheel)

As young designers become more mobile – and the opening up of Europe is simply another large step in a process that was already well-established in the 1970s and 1980s – they take their own cultural standards with them. In this fluid situation it will become ever more difficult to draw broad conclusions about a nation's psyche by looking at the designed objects it produces: each country will be seeking to export into an increasingly homogenized marketplace.

4 Travels in time

First-time visitors to Manhattan sometimes remark on the strange impact that its architecture has upon them. In the mid 1980s someone described it as looking like 'a 1940s vision of a futuristic city'. The interesting point about this observation is that it reveals two levels of understanding: firstly, it identifies a past 'vision of the future' – a built-in irony since any *accurate* past vision would look increasingly like the present as time progressed; secondly, the visual characteristics of that past vision seem to identify the period which produced it – for example, the Chrysler Building (1930), with its exuberant images of hubcaps and radiator ornaments, reflects a yearning for an automotive future that nevertheless seems firmly rooted in its time.

The style of many types of products is constantly invaded by visions of predicted futures and real or imaginary pasts. Such products also covertly identify the period in which they were designed. One could analyse a vast number of artefacts and speculate endlessly about their references, both intended and unconscious, to other periods in history; one could also examine them for what they reveal about their own times. What is more useful in the present context is to consider to what extent artful references to the imagery of other periods provide a way of using style successfully. First, however, it is helpful to take a broader view of stylistic 'travels in time'.

Past futures

It seems to be at those times when technological revolutions are imminent that the future exerts its most pervasive influence on the shape of designed objects. This is a slightly different process from simply designing for an abstract future, for it may be rooted in 'real' technology which has yet to express itself. This phenomenon occurs because it is in the nature of technological revolutions that the means frequently exist before the social and commercial initiatives which finally set them in motion; the result is an imminent world of tantalizing possibilities, glimpsed but not yet focused.

The scientists and designers responsible for the 1939 New York World Fair had anticipated a future in which nuclear power existed, but the shapes and images they created to celebrate the new force were necessarily imaginative rather than appropriate or accurate. However, whilst the architecture and some of the other visual conceits of the fair could never hope to be more than well-informed science fiction, the relentlessly futuristic imagery still managed to be both inspirational and influential. The New York fair was essentially a celebration of streamlining, and streamlining was a celebration of a future in which technology would show the way.

The seductive lure of the future was a phenomenon which would co-exist with other design trends for several years. It resurfaced in Britain in the 1950s with the Festival of Britain and a suddenly keen post-war appetite for symbols of a drudgery-free future.

The future will inevitably turn out to have different contours from those which we may project for it, and, even though it can stimulate influential and inspirational imagery such as that of the 1939 New York fair, it is likely to offer an unsure model for design. The Kolster

The Kolster Brandes FB10C radio draws on the imagery of science fiction in an attempt to appear futuristic

Brandes FB10C Radio dating from 1950 is an interesting example. A British bid to take radio into the future, it apparently took for its model the spaceship equipment in Universal's hilarious Flash Gordon cliffhangers. But this particular attempt to use the future as a source of successful stylistic imagery does not reflect a particularly sure touch. Although futuristic images often attract the consumer, they will not themselves guarantee a product's success unless proven appropriate by sufficient research and analysis into the target market.

Today's future

In dramatic contrast to most previous experience, the 1990s find technology for the first time outstripping both demand and popular comprehension. Never can there have been less motivation for futuristic styling in product design, since the overriding problem for consumers and industry is now trying to keep up with the implications of new technological advances.

The main purpose in setting up The Media Lab at the Massachusetts Institute of Technology was to provide some kind of major force for guiding new technology into useful channels – left to the technocrats it was exploding in all directions. This pioneering approach has since trickled down to bring about the establishment of other international bodies motivated by the latest technological revolutions, which now incessantly offer undreamed-of possibilities for dressing up what is essentially the same product in many different clothes (see Chapter 5).

The technological future now looks to be so exciting and unlimited as to have made inspirational imagery largely irrelevant and inadequate. Not surprisingly, it is no longer the future which excites and inspires stylistic borrowings, it is the past.

Times past

The past, like the present and unlike the future, exists concretely. To the designer it therefore presents a vast repository of imagery which can be alluded to, hijacked or reconstructed to suit the occasion. This process is in itself dynamic and subject to fashion. Perceptions of the present will determine exactly how the contours of the past are recycled. Selectivity is the key to success.

The first selective impulse comes in the type of product deemed appropriate for allusive imagery. When looking at child safety seats for cars, no-one expects to see amusing reworkings of the restraining equipment once used on the mentally ill. This is not an aspect of the

Pine and Victorian bathroom ranges for Marks & Spencer by Wharmby Associates: nostalgia carefully reinterpreted for a conservative contemporary market

Seymour Powell's Retro hairdryer for Clairol seeks originality through the subtle manipulation of historical associations

past we want to revisit. On the other hand, kitchenware and furnishings are susceptible to a popular nostalgia for the 'wholesome' values of the past. Shaker furniture is admired for its spare, clean lines and absence of ornament; solid Victorian kitchen utensils evoke an age of substance and quality. However, few contemporary consumers of products 'inspired' by these models would also subscribe to the religious ideals of the United Society of Believers in Christ's Second Appearance or the social policies of Victorian society. It is simply that certain seductive lifestyle associations have grown up with these types of products.

The second selective impulse comes with adjustment of the original model. Shaker-style cupboards are commercially available manufactured from medium density fibreboard which is first laminated and then distressed to mimic many years of virtuous use. Victorian-style utensils may be made out of lighter metals and be more elegantly finished in order to accommodate current notions of convenience and quality.

Designers may allude to the past through straightforward imitation or subtle hints. The Victorian and Pine bathroom ranges by Wharmby Associates tread a fine line. Designed for Marks & Spencer, they are required merely to hint at a nostalgic past. Neither playfully allusive nor aiming for 'reproduction' impact, they still aim to be taken seriously. As such they may be seen as representative of a silent majority in product design, where visual style is required to create an appropriate mood in an unspectacular way.

Imagery from the past may be recycled in a self-conscious way without endangering the elegance of the product. A natural consequence of German rationalism, the Retro hairdryer designed by Seymour Powell for Clairol is a response to the uniformity brought on by formalized shapes and simplicity of form and colour. When all contemporary hairdryers looked the same – featureless white or pastel plastic cases – the problem was how to differentiate one model from the rest. Seymour Powell's answer lay in appropriating and refining the distinctive imagery of another period, in this case the 1950s. The effect even includes a flecked cable.

There is something disarming about the Retro hairdryer in that it neither hints vaguely nor copies slavishly an historical original – but should one be suspicious of any piece of contemporary design which seeks to impress by bringing along historical baggage? Whereas the Modernist-versus-traditionalist argument in architecture is kept alive by the issue of how best to intersperse new buildings with old ones, there is no comparable issue in most product design. We may be inclined to mistrust any product designer who cannot find a distinctive form of contemporary expression for a product, but we do not live in a visually sterile world: many objects carry historical associations, particularly, for example, the reproduction furniture with which so many British people like to surround themselves. Manipulation of such associations is not only permissible, it is desirable unless we seek to live in a landscape of anonymous commodities where everything becomes neutral equipment. Somewhere between that arid vision and the necromancy of endlessly recycled Victoriana lies fruitful territory for the designer with a sure sense of style.

5 Shells and boxes

Any discussion about style in product design must at some stage approach the thorny question of where functional necessities end and stylistic possibilities begin. The next two chapters examine two contrasting aspects of the form versus function battle: firstly, products with the same basic components which can be presented in different forms; and, secondly, those highly engineered products which it seems place severe restrictions on any variation in form.

Since the 1920s there have always been products that lend themselves to being arbitrarily encased in shells or boxes. Quite frequently these shells and boxes have done little more than safely contain machinery or electrical components, revealing nothing about their workings or operation in the process, but more usually they have been used as convenient surfaces for expressing ideas about the products.

Streamlining

Raymond Loewy's redesigned office duplicator for Gestetner in 1929 converted a piece of messy industrial equipment into a formalized object, giving an early demonstration of the possibilities of style in product design. Meanwhile Harley Earl's automotive design work for

General Motors, starting with the La Salle in the late 1920s and proceeding through a whole range of fantasized Pontiacs, Chevrolets and Cadillacs, established the principles of styling: the application of a theatrical 'cover' to a functional machine for the sake of increased desirability – and therefore sales. Because initially the opportunity for such flourishes was that much greater in large objects, it was usually for these that streamlining principles were first established. Even so, when it became fashionable in automotive design, its eventual effect was to trickle down to smaller manufactured products.

Streamlining is important in illustrating how the same basic product can be adapted externally. For all its superficial claim to functionalism, it was by definition an exaggeration of aerodynamic principles. Aeroplanes might legitimately be designed to accommodate airflow, but streamlined automobiles elaborated upon the same imagery simply to create effect. By the time streamlining's imagery had been further adapted to toasters and refrigerators, any pretence at function had long since been abandoned. The legacy of streamlining was the principle of simply encasing objects in cases or shells which were expressive not of the way the products themselves functioned but of some vaguely abstract superimposed aesthetic.

New faces for old friends

A perfect contemporary demonstration of this enduring principle is the Philips clock radio by British designers Seymour Powell. In this case very sophisticated designers have accepted the principle that shells and cases frequently have little or no role to play in explaining how a product works. In common with innumerable other electronic products, the clock radio exists only as a few insignificant components too small to fill out a convenient-sized product profile. Emphasis is therefore placed not upon the function of the product but its image which is conveyed through visual style.

For this Philips clock radio Seymour Powell radically rehoused unchanged components to suit the visual expectations of a new generation of consumers

When clock radios first became popular in the 1960s, componentry was already small. The novelty of choosing whether to be woken by music or an alarm was sufficient to recommend the product to the first generation that bought it. As a result, form and visual language were of minor importance, and this was reflected in the Philips clock radio of the time. But what was then a technological novelty is now commonplace, and most households contain many far more elaborate electronic gadgets.

Seymour Powell concluded that it was therefore unreasonable to expect the product to go on recommending itself as it had originally done. Taking the same internal componentry, they devised a totally different shell for what was exactly the same product. The appearance of the new shell in no way makes excessive claims for the product – it does not try to look like more than it is – but there is a conscious attempt to place it more logically in the context of contemporary electronic goods.

The underlying principles of the clock radio also affect all manner of hi-fi and video products. When opened up, almost all of these reveal surprisingly underpopulated interiors. Much the same can be said

The Trimline Touchtone 1300; the Trimline was the first modern telephone to reunite microphone, earpiece and dialling mechanism in one neat sculptural unit

for telephones which, the more functions they incorporated, managed to compress componentry into a relatively small space. This has freed designers to try infinite variations on what can be done with a keypad and a handset, taking the emphasis away from the functional components and placing it on ergonomics and tactile values.

One of the earliest of the new generation of telephones came from the same New York consultancy that designed the famous Bell telephone of 1937, Henry Dreyfuss Associates. The Trimline phone, which subsequently sold more than 65 million sets, took a radical new step which was founded in ergonomics but which appealed through its imaginative visual and tactile style. Designed by Donald Genaro for AT&T, the Trimline was initially a response to consumer requirements for a compact phone. The old Bell telephone had always made rather less of itself than some of its European equivalents, and here, apparently, was a further wish from American end users to make the phone less obtrusive and easier to store in small places. The solution was to create an instrument which, when not in use, revealed no feature (except the cord) to identify it as a telephone at all. There is something rather witty about this – it goes further than the brief apparently demands, but it does so in a very rational way.

In the Trimline telephone the three basic phone functions –

microphone, earpiece and dialling mechanism – were reunited for the first time since the late nineteenth century. Integral with the handset of the Trimline was the dialling device (miniaturized rotary dial or push-buttons), so that when the handset was returned to the rest, all its working features were hidden beneath a curved surface shaped to fit comfortably into the hand.

The freedom of technology

The microchip has reduced the internal componentry of many products, sometimes leaving a wide range of options for the designer, who must then evaluate the visual message to be conveyed by an unconstrained external shape. The British company Waymaster once manufactured traditional weighing scales, where product design decisions were strongly informed by the nature of the product. A new electronic weighing mechanism developed by Waymaster in the UK suddenly meant that their scales need not be as large as before and new external forms became an option.

The Cardiff-based Nick Holland Design Group worked with Waymaster to create a scale based on the new weighing mechanism. It was clear that here was a chance to create a stylish new product. One option might have been to take the hi-fi route and create a shell or case which expressed ideas about the new machine. In fact Nick Holland devised a series of scales – kitchen, parcel, diet – which capitalized on the new miniature weighing device to

Freedom from the constraints of traditional mechanisms allowed Nick Holland Design to create a distinctively stylish range of scales for Waymaster

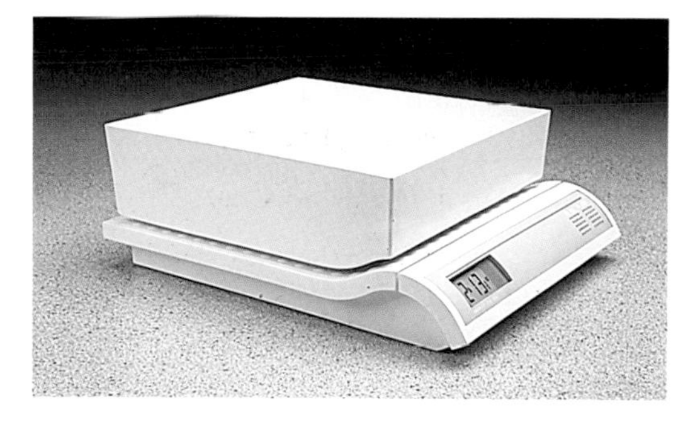

create a highly distinctive product clearly differentiated from the competition. The new Waymaster electronic kitchen scale works by means of an innovative load-sensing system that guarantees good load stability and a high degree of accuracy on any part of the platform. The opportunity to display the smallness of electronic componentry in non-portable equipment is often rejected in favour of large empty cabinets, but here that opportunity is well taken: the wafer-thin profile means that the most prominent facet of the machine is that which performs the most important function, the weighing platform.

Approachable technology

Unlike consumer goods, industrial machines and medical equipment might not at first glance seem to have the same need to be persuasive or aesthetically pleasing. Reliability, price, ergonomic performance and safety may seem to be the determining factors in most cases. For many years that was often true – witness the ugly machinery that dominated most manufacturing industries in the 1950s, and witness too the grotesque 'industrial' appearance of many pieces of medical equipment still in use in hospitals.

In fact there is enormous potential for using visual style to positive effect in both of these areas. Whilst some heavy industrial equipment still works best on a form-follows-function basis, the rise of the computer and a wealth of high-tech equipment has encouraged a reassessment of how we treat these professional products. Faced with the prospect of purchasing equipment, few committees or executives can claim to understand fully the way it works. Old proprietorial attitudes held by the purchasing company's senior employees, who

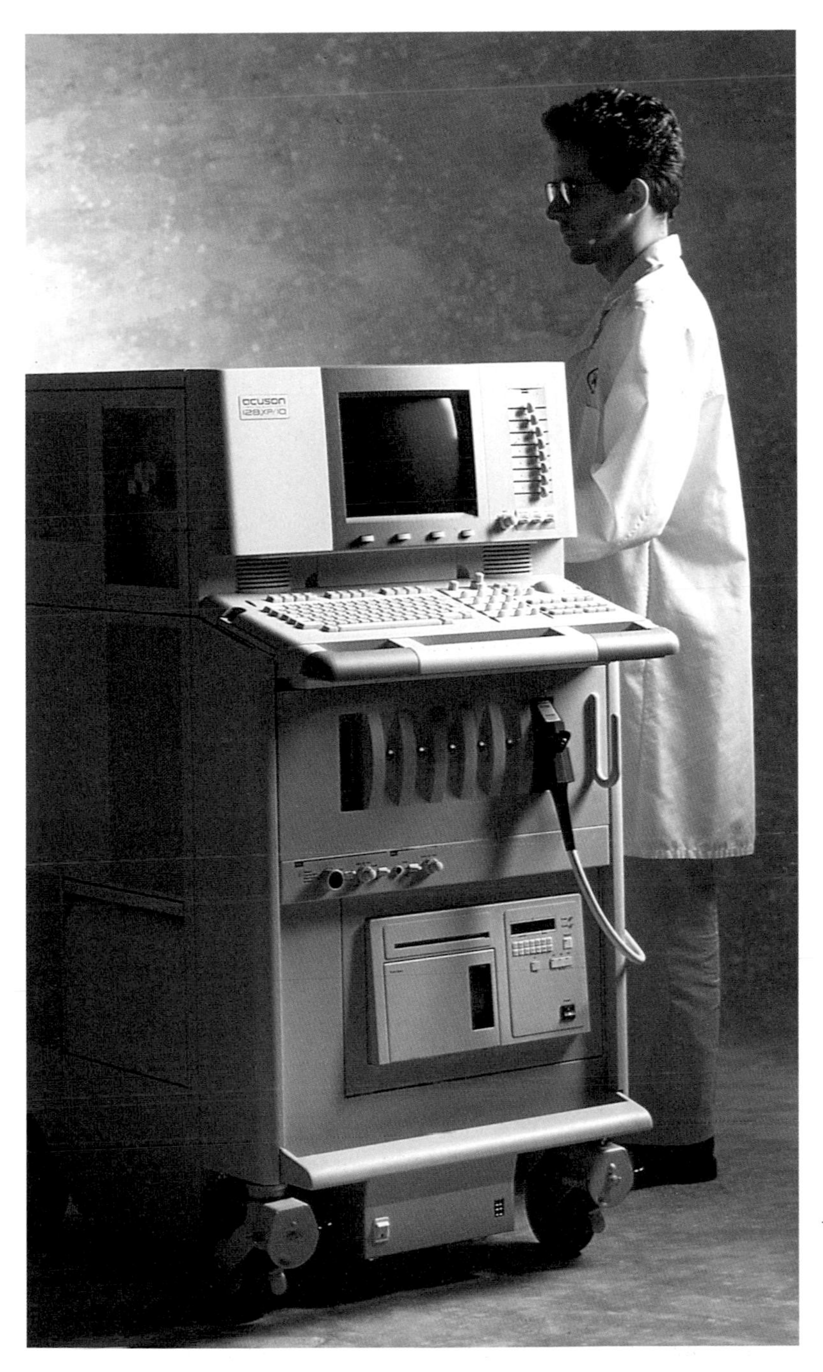

Medical equipment can be made more user-friendly and comprehensible through style: the Acuson Medical Ultrasound Imaging System

often had detailed knowledge and accompanying prejudices, have been replaced with a new type of business thinking. After identifying what tasks a particular piece of equipment is capable of performing, the potential commercial purchaser may be confronted by possibilities from several rival manufacturers. On what basis is a decision then going to be made? Inevitably the consumer factor reappears. Which one 'looks best'? Which is 'the nicest'? Visual style is then the determinant.

Improved design standards in medical equipment can bring many benefits, not the least of which is making it easier and simpler to operate. IDEO's work in this field has demonstrated some of the advantages to be gained from using the surfaces of medical machines to project something of their function. The Acuson Medical Ultrasound Imaging System (see previous page) is a large piece of equipment whose appearance threatened to become as unwieldy as its name. In redesigning it, IDEO sought to improve the appearance and signal the sophistication of the system. The importance of controlling details on a large piece of equipment is also well addressed.

One of the effects of improved medical treatment is the increasing availability of medical equipment which almost takes on the role of familiar 'accessory' for the patient. In such cases there is a need to make products not only safe and simple to use, but also user-friendly. Dialysis machines should not accentuate any sense of isolation already felt by a patient receiving treatment. Equipment which is actually worn has an even greater psychological need to be perceived as sympathetic. Any variety of cases or shells might be used to house such equipment, but only something which is sympathetic in both visual and tactile terms can be considered to have achieved successful style.

The Baxter Ventricular Assist System illustrates this point. The system is a power and diagnostic control unit worn by heart failure sufferers during preparation for surgery. Designed by IDEO, its organic, flowing surface makes a potentially hostile and invasive piece of equipment far less threatening and more comfortable.

Pragmatism for profit

One of the most ingenious extensions of the 'shells and cases' approach was implemented by designers Seymour Powell in the late 1980s. Upon establishing a relationship with the new owner of the British motorcycle company, Norton, Seymour Powell were anxious to embark upon designing a new motorcycle for what was by then an almost forgotten British industry. Before they could do so, commercial considerations required that they address an existing Norton machine which could not be fundamentally altered. The solution was to examine the small existing market for the machine, which largely consisted of export orders for paramilitary use for which the machine's modest but

Norton F1 motorcycle designed by Seymour Powell; the company's work for Norton initially involved unifying the surface appearance of an existing product and integrating the most commonly required equipment

reliable performance was no hindrance: it would be decked out with panniers, an extra battery to power auxiliary attachments, and other pieces of equipment suitable for this sort of application.

Seymour Powell identified that with such limited options open to them the best solution would be to unify the surface appearance of the machine whilst increasing its appeal by integrating all of the most commonly required equipment. It was an undeniable styling job, but one undertaken not for meaningless aggrandizement but for commercial sense. It also aimed to provide a financial opening into a more adventurous subsequent programme.

1950s 'styling' in the American sense now looks vulgar or comic. Tiny clusters of electronic componentry housed in almost-empty boxes may to some suggest a regrettable move towards arbitrary product design. And yet good designers are managing to refine the underlying impulses of both of these developments to create products with acceptable 'style'. Casings which express what a machine has to do, or which make it more inviting, are simply reworking the fundamental principle of the first American styling, albeit more thoughtfully. Tiny electronic components also liberate the designer to make creative statements which need in no way be theatrical or exaggerated. In turn these exercises stimulate designers to reconsider the exterior treatment of products which – like the Norton motorcycle – may, perhaps unexpectedly, be enhanced by an intelligent 'superficial' treatment.

6 Form, style and engineering

In clear contrast to the principles discussed in the previous chapter are those governing 'intolerant' or highly engineered products. If the product has to perform very precisely, if stress and weight factors are crucial, if altering the form significantly will undermine its performance . . . can it really be 'designed' at all, let alone stylish or aesthetically pleasing? Is this not in fact the realm of engineering design, a more organizational process concerned with function and efficiency?

The art of engineering

One classic model sometimes used to define the process of design is that of a commercial aircraft. Balancing a whole set of conflicting practical demands results in the design of something which still manages to get aloft despite having to accommodate passengers, luggage, fuel and a whole array of other things inimical to flight. If an aeroplane's design were left to the purist aeronautics engineer it would probably look like a glider. Left to the airline accountant it might resemble an ocean liner with wings. Both solutions would be equally impractical, but a well-designed aircraft satisfies both the needs of commerce and the laws of aerodynamics. There is, however, very little room for design manoeuvre, which is why aeroplanes all look basically rather similar – distinctions are mainly determined by capacity and engine type. Gliders meanwhile tend to be even less open to variation than other types of aeroplanes and usually manage to achieve a kind of artless beauty leading one to assume that it is almost impossible to design an ugly glider.

The same might be said of bridges. The best ones have a look of inevitable beauty and minimalism which somehow suggests that they could not really look any other way. However, it is still perfectly possible to build an ugly bridge, since brute strength will also do the job in many cases.

Aircraft and bridges point up a useful principle: where there are strict limitations imposed by physical function, there is often an apparently severe constraint on the design option. But it is just such constraints which can sometimes lead to the kind of rigorous design that embodies a certain highly satisfying form of visual allure – style seemingly born out of intractability.

An examination of three basic and very familiar items – a bicycle, a table, and a watch – shows how this principle can in varying degrees be relevant to many other, smaller objects. The designer may ask a range of questions when approaching a new project. Is the new solution going to be change for change's sake? Has the 'familiar' product actually undergone a subtle shift of role over a period of time, so demanding a new treatment? Or has the need for real design improvement remained largely unmet because of the overwhelming success of the original design? The extent to which style forms part of such solutions is likely to derive from the imaginative quality of the response to such challenges.

A bicycle

The bicycle remains an extraordinary and seductive challenge to the designer. Architects in particular are attracted by the idea of something that looks so simple but which incorporates a surprisingly large number of engineering feats. The ongoing bid to redesign the bicycle is a rich source of unintentional comedy and rarely achieves even qualified success, but it is highly instructive to consider some of the more legitimate design responses to the challenge.

Since a bicycle has to perform to a certain level, and since its performance is closely related to the expending of human energy, everything about it must be designed to facilitate power transference, comfort and resilience. Most alternative designs fail at once, because they have made the fundamental error of embarking upon a radically alternative design without fully understanding what the product has to do. By the time the shortcomings have become apparent the designer may be too committed to the new solution and wastes even more time tweaking a fundamentally flawed concept.

The Moulton separable small-wheel bicycle contrasts greatly with such failures. Alex Moulton, who designed the suspension system for the Mini, has come to typify the obsessive bicycle redesigner. This particular solution is a triumph of will over common sense: having set himself an elaborate, interconnected set of problems by insisting on small wheels and a separable frame, Moulton then solved these problems quite brilliantly. The Moulton space frame bicycle is a superb piece of engineering resulting in a machine with a very high price and only slightly reduced performance compared with the conventional design.

Broadly speaking, no-one has yet managed to improve upon the basic design of the 'safety' bicycle, first introduced in the late nineteenth century. Indeed, the mountain bicycle, which has gained worldwide popularity in the consumer market in recent years, resembles the original safety bicycle even more closely than it resembles racing bicycles, for which modified handlebars and pared down components are the norm.

This is true of titanium-framed mountain bikes produced today, which embody most of the new 'design' elements of the 1980s: improved alloy construction, more sophisticated transmission, gear and braking components, and a return to the straight handlebar configuration of the original 'safety' bike. In fact the materials from which frame and component are made are all that have changed. Improved alloys and carbon fibre have been introduced for weight saving, and numerous minor improvements in mechanical linkages have taken place. Yet, surprisingly, bicycle technology has provided the basis for an extraordinary stylistic development that has resulted in a major shift of manufacturing dominance.

A crucial range of products make up the transmission, gear-changing and braking components of the bicycle. These 'groupsets', as they are known, are marketed as product ranges in their own right since bicycles, unlike cars, are usually assembled by distributors or dealers from an informed choice of frame, groupset and other interchangeable parts. In the post-war period an Italian company, Campagnolo, dominated the upper end of the global consumer market for bicycle componentry. With no perceived possibility of raising design standards, Campagnolo dominated through a blend of prestigious reputation and high quality materials and bearings.

When a challenge came to Campagnolo, it came from Japan.

Shimano had been making bicycle components – amongst other things – from the 1920s onwards. In the 1970s they identified a US-based boom in cycling and began to address the burgeoning market. Today Shimano has effectively overtaken Campagnolo in the world market by producing many ranges of groupsets that make no concession to cosmetic design – there is simply no room for superfluous embellishment or impractical features in products which are used in competition – but which project a strong image of Shimano style.

Shimano's remarkable achievement can be fully understood only by appreciating the rigour of the engineering design that shapes such products. Ever greater levels of performance are demanded and must be satisfied, and Shimano has been single-minded in pioneering technical improvement of virtually every lever, sprocket, arm and mechanism within the various groupset. The company's product ranges, with their beautifully machined finishes, subtle contours and sophisticated packaging, nevertheless have a powerful visual appeal to the consumer. As a result the bicycle components produced by Shimano manage to project an image of distinctively engineered jewellery.

Shimano's jewel-like bicycle components manage to look alluring despite the rigorous constraints imposed by function and performance

A table

A table appears to offer comparatively little opportunity to the designer to improve upon the familiar arrangement, although it certainly offers more scope than a bicycle. Common sense and centuries of empirical research suggest that it will probably work best if it is rectilinear and has a leg at each corner. Embellishment upon this simple scheme must necessarily be superficial; or at least, so one might think.

The Nomos 'Father' table, designed by architect Norman Foster for Tecno, combines steel, aluminium and glass to form an elegant and practical 'architectural' solution to designing a prosaic object. The result is a triumphant exercise in minimalistic style rarely found in mass-produced furniture. Certainly it cannot improve upon traditional stability and performance. Equally it is very much more expensive than many alternatives which perform just as well. Yet in seeking a new form for a traditional product Foster has actually arrived at something which

Stylistic originality with a purpose: Tecno's Nomos Father table by Norman Foster brings added value to the familiar concept of a table

has both a distinctive style and a legitimate claim to being more flexible in application than most tables. Foster envisaged his table as being equally suitable for conferences or dining. By applying the kind of technological thinking developed for innovative architectural solutions, arrived at a highly functional table which looks so unlike traditional tables that it is equally comfortable in a social or business context. In addressing an apparently intractable problem Foster arrived at a satisfying product with a clear style of its own.

A watch

Watches are traditionally valuable items, miracles of mechanical miniaturization. The constraints of size, weight and legibility meant that originally only invisible jewelled movements and the introduction of precious metals – gold or platinum cases – offered opportunities for design variation. Later, elaborate extra features were occasionally added, particularly to men's watches, converting them into expensive pieces of pseudo-military equipment which recalled their image-forming role as functional aids for German gunners in the First World War. For example, Louis Cartier's 'Tank' watch, incorporated mock military imagery and had already abandoned numerals as early as 1917.

With the advent of the microchip, digital watches arrived. More robust and in the end so cheap as to be almost disposable, they latterly invited the more playful attentions of designers. The demands of size and a degree of practicality still meant there was little to work with, but this encouraged ever more outrageous solutions in the popular market and 'style' – in its fashionable sense – became the essence of Swatch products. Intrinsic value is no longer a prime consideration.

M & Co watches of the late 1980s proved how experimentation could create playful, anarchic or outrageous styles. Today the restrictions of size and legibility are still essentially unchanged, although it is interesting to note that whilst the digital watch can be very accurate, few watches any longer bother to calibrate the face, depending instead on a notional reading.

It is as interesting as it is unexpected to find some of the most persuasive examples of successful style appearing in products whose nature might seem to exclude all but the most functional approaches.

7 'Special' needs

Putting aside seductive imagery drawn from different times and different cultures, and stylistic features influenced by functional restraints, what are we left with? What other kinds of aspirational visual or tactile style can be imparted to products to make them desirable – or is mere 'desirability' missing one of the most important points?

It is likely that in the future one legitimate role for style in product design will be to address more fundamentally humanistic needs. There are of course numerous genuine 'special' needs, but 'special' may sometimes blind designers and manufacturers to the fact that needs are not always as different as they seem. Careless assumptions must be avoided if truly appropriate designs for special groups are to be created.

Green design

Many of us might believe that a world in which environmentally responsible behaviour encompasses all aspects of design, manufacture and marketing is desirable. But, equally, environmentally responsible products or services may need to project their virtue with the help of the same styling techniques as their competitors: they may be socially responsible but they may nevertheless still need to look desirable.

In October 1991 an early bid by the Design Business Association

in London to address these matters resulted in a modest exhibition of prototypes from four British consultancies: HALL Richards, Grey Matter, IDEO and Somerfield Design. The aim was to show manufacturers and consumers that green machines need not look ugly. As Vicky Sargent of the Design Business Association remarked of green products, 'They don't have to have a hair shirt look about them – they can be made as desirable as any other consumer durable'.

Grey Matter's can crusher is proof that green products need not have a hair shirt look about them

Environmental demands highlight two other important points: that fulfilling a need is in itself not enough – products have to be *seen* to fulfil the need; and that products for special needs have an important place in the style discussion.

Packaged goods like detergents at least have the option of flagging their 'greenness' (or otherwise) in a straightforward way, but products need to project their greenness through their form. As with most new concepts, to begin with there seem to be no obvious aesthetic means of achieving the desired effect. Greenness will no doubt find an eventual visual vocabulary of its own, but for the present even the simple tree, cloud or leaf imagery of graphic design seems to be more advanced than most product design in this field. This is not necessarily due to lack of imagination amongst designers but to caution from manufacturers, the longer lead times involved in product design and the far greater expense involved.

Two product ideas from IDEO show how the needs of green products can be satisfied. In the Bin Bank, which can be made from recycled polypropylene, a rotating carousel separates cans and different coloured bottles. The kettle has an upper water reservoir and a lower boiling chamber which admits precisely metered amounts of water for energy saving. Sound ideas are self-consciously presented as looking 'different'. The Dali-esque kettle looks almost as if it may be melting with the burden of all that saved heat. Meanwhile the bin lid, with its hint of yin and yang, also seems to be trying to tell us something.

Bin Bank and kettle prototypes from IDEO: environment-friendly products designed to signal their social attitude

Choice for a changing society

It is difficult to deny that design should provide the right amount of choice to all members of society. In the post-war years it has been the youth market which has dominated many areas of design, but the lack of well-designed products for people aged over 50 is now attracting serious attention from certain Japanese, American and Scandinavian companies. A research programme, DesignAge, at the Royal College of Art in London has also been established to help change a situation in which, since the mid-1950s, industry and commerce have been almost exclusively hypnotized by the needs of youth. It is estimated that early in the twenty-first century there will be more than 130 million over-fifties in the European Community alone.

In many cases there is no reason to suppose that the function of products will necessarily need to change. Rather it will be the style. There is nothing odd about a middle-aged adult wanting to purchase a portable radio-cassette machine to listen to music or radio plays about the house; but it is odd when that potential consumer is confronted with a range of extrovert products aimed solely at young people.

In other cases, and by no means exclusively in the over-fifties age group, there is a genuine need for products that can be used by people with some disability. To design only 'special needs' products, where the gesture of producing them at all is apparently seen to be sufficient to absolve the manufacturer from all obligation to make them attractive, is an act of social condescension. It is always possible to bring the same levels of visual quality to such products and it is frequently possible to blur the distinction between 'special' needs and ordinary ones.

An excellent example of this is the Good Grips range of kitchen gadgets from the New York consultancy Smart Design. As the

The style of Smart Design's Good Grips kitchen equipment for people with impaired dexterity extends the range's appeal to other consumers

designers claim, these products 'demonstrate that ergonomics aren't just for automobiles or medical equipment and that transgenerational design doesn't have to look frumpy'. Working on the principle that people with special needs should not have to search for special products, Smart have designed many of these tools with large handles covered in thermoplastic rubber. The finned soft spots on the handles are enjoyable to use and prevent the tools from slipping out of wet hands. Their practical advantage is that they can be used across a wide range of dexterity levels.

Measuring cups and spoons in the range also have extra large handles for better control, and feature colour-coded denominations. Tools which need to be squeezed – scissors and can-openers – have added leverage and wide flat handles to minimize pressure on the hands of the user. The simple result is that these products are easier for everyone to use and, in addressing ergonomic issues, they have made a visual (and tactile) virtue out of their practicality. They extend the range of potential users whilst making a bold visual statement about their intentions.

Design for children

Unlike other groups in society, children are frequently either ignored or treated as a race apart. As such they frequently become the recipients of inept and offensive products which are designed and dressed up to appeal to them by adults whose understanding of their requirements is grotesquely misinformed. There is a strong case for arguing that of all groups disenfranchised by poor design, children are the most important for the most obvious reason: it is not only their immediate environment which is being affected, it is also their psychological development and their future tastes.

Nasta Industries' Blabbermouth cassette player; children's products are sometimes too influenced by adult perspectives

In the USA one of the most perceptive and energetic crusades against the muddled intentions and simple vulgarity of much American design for children is the non-profit, New York based CHILDESIGN. Taking a broad cultural attitude to the issue, the organization's stated goal is 'to create a place of value for children in the culture at large'. From a whole range of badly designed children's products in the USA, some can be seen to be manifestly bad by more or less any standards. Others fail when inappropriate visual style is applied to an otherwise adequate product. CHILDESIGN principal Sandra Edwards argues that design for children is often seen in terms of surface decoration.

'We don't really design for children', she observes, 'but in service to some memory or notion of childhood, without actual observation of children today'.

Products for children are sometimes too influenced by adult perspectives. They are also often unnecessarily vulgar, revealing misconceptions of what appeals to children. For some people Nasta Industries' Blabbermouth tape recorder typifies this kind of patronizing design thinking. Another of this company's children's products seems to rival the Blabbermouth for insensitivity: the Money Talks tape recorder comes in the guise of a large dollar bill. The joke – if it is a joke – is an adult one, the underlying assumption seems vulgar and the resulting visual style therefore almost entirely meaningless.

Such products may exemplify what is wrong with American design for children but, as an analogue of the way objects are designed for adults, design for children can be highly illuminating. By making unfounded and uninformed assumptions about children's needs, designers can unwittingly perpetuate myths and impose strictures upon the end user's environment. This frequently happens with design for adults too, although in that case there is usually more choice and the consumer is buying for him- or herself or another adult. When it comes to buying for children, the adult is likely to be swayed by a bold visual statement, which may be entirely inappropriate but which somehow satisfies a vague feeling about what is suitable for children.

Sandra Edwards explains that there is a fundamental problem with graphics: they are rarely relevant to the product. There is also a styling misconception about colour: 'Developmental research on visual acuity shows that infants cannot distinguish between subtle colours. But manufacturers have overgeneralized this finding and red, blue and yellow have become synonymous with childhood. When applied to objects with long-term use such as furniture, these colours work against flexibility and leave little opportunity to alter the environment to match the growth or changing moods of the child'.

Products for real needs

Whilst many products, including those for the seriously disabled, demand major modification, there is no reason for their appearance to stress the isolation of people who may already feel isolated – quite the reverse in fact. Moreover, there is a major obligation on industry and designers to provide more thoughtful and humanistic design solutions in many areas. As ever, this can only be achieved by addressing real needs and requirements, including psychological and emotional ones.

Ironically, in some of the above cases designers may be overtaken by events. The urgent practical necessity of greater environmental concern may propel us into buying green products without allowing time for the development of a visual vernacular to persuade us to use them. The increasing spending power of the over-fifties may itself accelerate the development of products more visually and practically suited to older consumers. A generation of computer-literate children, who are frequently the only members of the household able to operate the timer on the video, is unlikely to perpetuate the muddled visual legacy of 'imitation' primary-coloured TVs and radios for its own children. Even so, the development of valuable stylistic language for product designs with a social dimension remains one of the more legitimate challenges to the design profession.

8 Successful Style

In this final chapter it is appropriate to offer some examples of products which many of us would agree are highly successful stylistically, and which not only appeal and attract by using the principles discussed in this book, but also achieve a great deal more. In each case the design approach and what was learnt in the process will prove illuminating.

The T3 soap dispenser and washstand

The London-based multidisciplinary design company DEGW has a product arm called T3. Working in the relatively sheltered world of creating products for architectural specifiers, T3 is able to pursue a policy of excellence in an area where cheap standard solutions are the norm. The consultancy tends to favour a metaphorical approach to devising products and, on being asked to develop office washroom equipment that combined a high quality image with advanced standards of hygiene, it assessed the options by sometimes fairly oblique means.

In searching for a form and materials to create a small initial product, a liquid soap dispenser, T3 noted that conventional solutions inevitably connoted cheapness – usually through using plastic

Washstand designed by T3: visual elegance and sophisticated technical operation

components – and messiness because of the way they operated. One alternative image was suggested by a piece of veterinary equipment – a satisfyingly manufactured, pistol-grip, metal syringe powered by a Sparklets cartridge. Rich blue storage bottles suggested another pleasing material that might be deployed. Eventually it was the traditional imagery of the barber's shop that prompted a solution. Materials like ceramics, chrome-plated brass, vitreous enamel and natural stone proved to be consonant with the desired high quality environment because of their historical association with washrooms. The T3 soap dispenser, with its legible operation and high quality ceramic finish, achieves a 'professional' look without resorting to pretension.

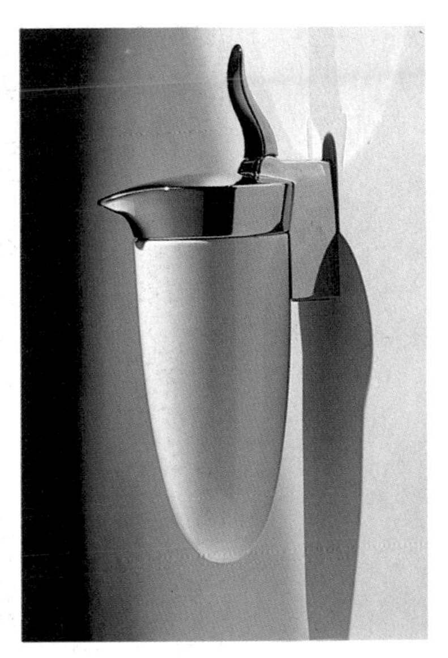

T3's soap dispenser combines legible operation, a high quality finish and 'professional' imagery

T3 subsequently created a complete washstand, remarkable because its sophisticated technical componentry could easily have resulted in a hostile appearance either openly expressed or thinly disguised. In fact the T3 solution makes high technical performance and ease of installation invisible attributes. The result is a piece of expressive design in which primary visual elegance combines with the technical features of touch-free control of all functions, ultra-violet prevention of bacteriological growth in the water system, a sealed soap supply system and an integral hand-drying facility.

Although neither soap dispenser nor washstand are in production at the time of writing, the way in which T3 resolved cultural, practical and financial issues in designing them shows how, no matter what difficulties the task presents, a stylish solution can be found.

The BIB Agenda

Electronic organizers are still relatively new concepts. The product they were intended to supersede – the Filofax – took the familiar form of a book and therefore never faced the issue of how it ought to look. 'Electronic books', however, face the now familiar problem posed by micro-technology: when the product does not have to look like anything in particular, what should it look like? Rival products from Psion chose to solve the problem by putting the calculator in a hard shell holster. Sharp decided that the model of a wallet or billfold was preferable. For the Agenda though, BIB Design Consultants aimed for a more imaginative organizational approach. Despite the multiplicity of buttons, the design's organic arrangement of functions is very clear and inviting; it also reflects something of the product's unique feature – Microwriting, a form of making characters on the LCD screen which corresponds more to natural writing than the simple punching in of discrete symbols.

BIB's prize-winning design resolved a difficult organizational problem with flair. Its appearance invites the user to investigate it – no small feat, since this particular product actually incorporates more features and buttons than some of its rivals. The key to its appeal is the almost informal arrangement of the controls. Looking like no other electronic product, it manages to look like a legitimate product in its own right.

Iroqua seating

The Iroqua series of detachable seats is by Catalan designer Josep Lluscà. It is intended to be appropriate for a wide range of situations, an objective which in many hands might have resulted in a 'lowest common denominator' solution. Instead the Iroqua series combines a remarkable number of practical considerations to do with ergonomics, maintenance, storage, transport and export.

The Iroqua sofa has a metal skeleton coated with injected polyurethane foam for the seat, back and armrests. Each section is independently constructed, and when assembled the components are not in direct contact, but slightly separated to prevent the usual build-up of dust and detritus in sofas. This type of construction also greatly simplifies storage and makes shipping easier.

Intended for use in the home, waiting-room or office, the Iroqua seating series is also specified for locations like banks, hotels and other public spaces. The aim of arriving at a visual style that does not look out of place in any of these situations has been remarkably well achieved. However, this is no simple exercise in visual style. Comfort is given high priority and the ergonomic principles employed are based on research in automotive and commercial aircraft design. Four different versions are available: low, medium, high and high with headrest.

Iroqua seating by Josep Lluscà: seductive style creating functional seating suitable for a variety of environments

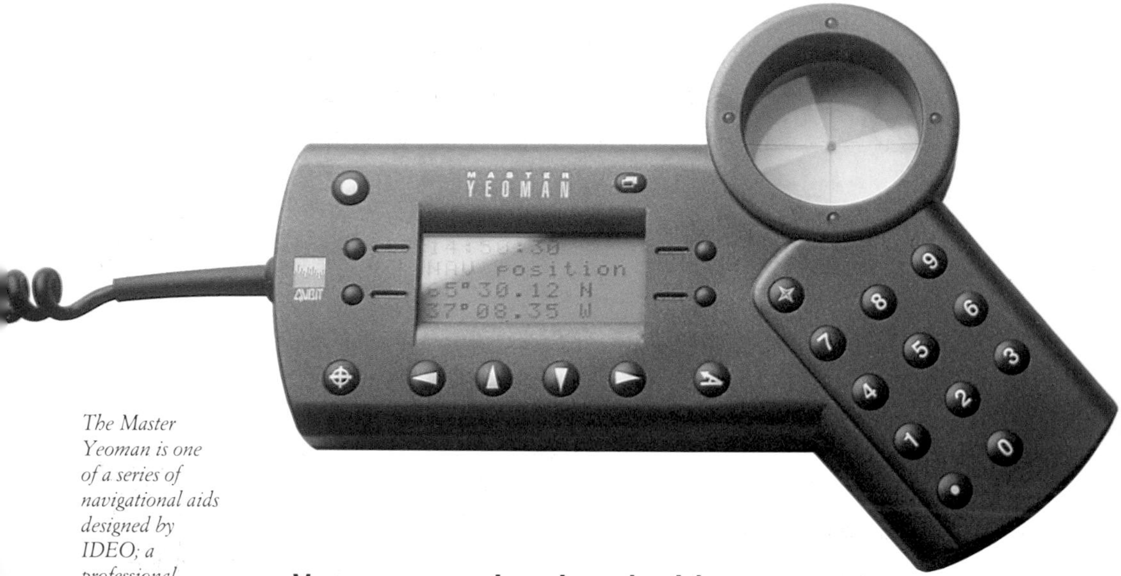

The Master Yeoman is one of a series of navigational aids designed by IDEO; a professional, precise feel is combined with a practical and accessible form

Yeoman navigational aids

The Yeoman series of navigational aids from IDEO represents an enjoyable progression of visual style from consumer to professional markets. The small leisure craft market was addressed first with the Navigator's Yeoman. The device allows for accurate navigational calculations and has very carefully married a professional feel for precision with a form that is both practical and accessible to the non-professional. Using the same technology the Master Yeoman was more sophisticated in its operation but brought the same seductive sense of style to an instrument intended to be used by larger ships such as oil tankers.

A nice inversion of the same idea was aimed at an even more specific professional group – cartographers. The TDS Numonics precision cursor further builds upon easy-to-use-but-professional imagery to facilitate the task of map-making, rather than map-reading.

David Mellor Provençal cutlery

Although the notion of designer-as-artist has been resisted throughout this book, David Mellor's 1973 Provençal cutlery, a supreme example of visual style, comes close to the realm of art and certainly deserves mention.

Design bordering on art: David Mellor's elegant Provençal cutlery

Mellor's first range of cutlery won a Design Council award in 1957. He has since brought a potent blend of rigour and elegance to many other products. The exact reason for the appeal of the handfinished Provençal range is difficult to pin down: there is a sculptural precision about the blades and tines, while the original rosewood handles (which were later replaced by resin) added an extra tactile quality. Countless knives and forks have been fashioned out of similar materials, but the visual and sculptural refinement of Mellor's Provençal range is almost the sole – and at the same time highly successful – distinguishing feature.

Visually seductive and invitingly tactile, Henry Dreyfuss Associates' baitcasting reels for Abu Garcia improved ergonomic efficiency whilst creating a clearly identifiable new look

Abu Garcia 800 baitcasting reels

Henry Dreyfuss Associates can claim one of the world's most enduring reputations for product and industrial design in the United States. There is a consistency about its output despite the fact that it has been in business since 1929 and its founder died in 1972. With a reputation for ergonomic excellence, Dreyfuss is still capable of producing individual products that succeed as much through style as ease of use. An excellent example is the Abu Garcia 800 series baitcasting reels for anglers.

Abu Garcia had fallen behind the Japanese competition – notably Shimano – in US sales of its angling products. In fact the products were perfectly sound, but they lacked visual appeal compared with the Japanese rivals. Dreyfuss undertook the redesign of an existing product line, addressing a range of issues including user control and comfort as well as appearance.

The result, the Baitcasting Reel 821, was an exceptionally successful blend of features. The selective use of aluminium frames and graphite composite shells meant that the reels could be made both economical and tough. The new construction also prompted a new feature – a unique foldaway captive side piece which enabled rapid spool changes without the usual loose parts. The reels also feature a magnetic spool brake and the whole thing has a contoured finish for fitting into the palm. Even so the product range's appeal remains primarily visual. At the point of sale it is the seductive looks which first appeal, followed by an inviting tactile quality. The extra features – which perhaps represent the greatest value after purchase – are then revealed.

The AT&T Safari Notebook Computer

Another Henry Dreyfuss product in a quite different register is the AT&T Safari Notebook Computer. Here the objective is to create a portable personal computer expressly intended for sales and services personnel who work in the field. As a technical exercise it is a fairly simple matter to compress the required technology into a portable package. There are therefore many products in this field which offer similar features at a similar weight and size. However, the common solution is to create a white or beige box – a visual derivative of the most common colours for desktop computers – and to add a handle. The Safari makes the visual leap to the imagery of expensive luggage –

The AT&T Safari Notebook Computer draws on the imagery of high quality luggage to create a strong individual identity in a crowded market

which, in a way, is exactly what it is. At seven pounds it is not light, but at least its visual mass is broken up by dividing the form with detail and colour, and the ribbing provides a surface which responds well to handling and transportation. A detachable carrying handle can be used to tilt the computer into a position suitable for typing.

The Safari is an excellent example of the power of visual and tactile style. The product is in no way radical. The resolution of the detail is good but not exceptional. What really makes it a desirable object is the way it appropriates imagery from a different but appropriate product area and converts it to the job in hand: we are immediately persuaded by the high quality luggage image conferred by deep grey body colour, rounded corners and ribbed finish.

The Artemide Tizio task light

Richard Sapper's well-known 1972 reworking of George Cawardine's 1934 concept for an infinitely adjustable task lamp is an almost perfectly realized idea. The Tizio light for Artemide achieves a highly satisfying profile which explains how it will work even before you touch it. The necessary untidiness of the Anglepoise's wire was dispensed with by conducting the low-voltage current to the lamp through the structural arms. Meanwhile a transformer took the place of the originally weighted base.

Not so much invention as an imaginative deployment of newly-available technology, the Tizio lamp epitomizes the visual appeal of spare technology. The first time one sees it there is a pleasurable little start of realization at how it must work. Next comes tactile invitation to explore whether the counterbalances actually work and keep the lamp in any set position – and they do.

The Braun ET44 electronic calculator

When they first appeared, electronic calculators were remarkable novelties. They distilled the original purpose of the computer, an invention which was first conceived simply as a number cruncher, into an almost schematic little product. Since then innumerable pocket calculators have been produced, but the Braun ET44, designed by Dietrich Lubs and Dieter Rams in 1977, represents the most desirable version imaginable. Austere and yet playful, it is actually shaped like a pocket and the yellow 'equals' button stands out in unexpected contrast to the rest of the product which apart from this is in two shades of grey. It is difficult to imagine a more assured solution.

'Simple is better than complicated. Quiet is better than confusion. Unobtrusive is better than exciting. Small is better than large. Plain is better than coloured': Dieter Rams' view of design is expressed in suitably unembellished terms. Perhaps the ET44's riotous yellow button was inserted at the insistence of Rams' fellow designer Dietrich Lubs.

• • •

Although the preceding selection is a motley one its diversity endorses the theory that style has less to do with some established visual code or fashionable consensus of what looks 'right' – at its best it is more elusive and more potent than that. All of the products featuring in this chapter use style in an acceptable and satisfying way. Nearly all of them make a spontaneous and positive visual appeal; they are not casually approached and then admired, rather they attract attention and invite further examination. Attracting attention is in itself not that much of a trick – but then to sustain that attention and support it with satisfying revelations about feel, form or performance . . . that, it seems, is the very essence of successful style in product design.

Conclusion

Design, like any creative activity, cannot be learnt directly. The learning process is usually more a question of guided exposure to the work of others and analysis of the circumstances which brought it about. The same goes for style, whether in products or any other design discipline.

As I suggested at the outset, there is danger in focusing exclusively on the appearance of individual products for enlightenment and inspiration – although admittedly in certain cases it is hard to see exactly where else to look for explanations as to why a particular commodity is successful. (David Mellor's Provençal cutlery, for example, may well be inspired by some original regional model, but it is doubtful whether this is why it has sold so well over the years: it simply looks and feels good.) However, this merely confirms that design is not always neatly classifiable, and the appearance of the occasional inexplicable success is an important part of the allure of creating and shaping objects which people need, want . . . or are seduced into wanting.

Broadly we have seen that there are areas of influence and context which are commonly used to confer style, to stimulate benign associations and to make sometimes unexpected connections. 'Cultural context' can be simply paraphrased as inviting the designer to keep feeling the pulse of the times in which he or she lives. There exists today a democracy of popular culture which informs most aspects of our lives, and designers ignore it at their peril. Perhaps its most important characteristics are speed – speed of accessibility (TV, satellite communication, fax, cell phones and the rest) and an increasingly pan-global homogeneity.

The growth of multinational companies suggests further reasons for the erosion of clearly identifiable national market tastes. Not only are companies like Canon, IBM and Apple potent presences in the global marketplace, they are increasingly being encouraged to organize their research and design on international, or at least continental, lines. Media Lab in Cambridge, Massachusetts has major international sponsors who are constantly invited to witness the pre-production stage progress of various programmes aimed at shaping the way new technology is used.

Learning to encase products in housings which express something benign about them is a process which, as we have seen, has made great strides from the days of classic American styling; however, the temptation to impose irrelevant theatrical treatments remains. Meanwhile designers continue to rise to the perennial challenge of the 'unimprovable' product, with results which are sometimes unexpectedly heartening.

Environmental consciousness seems likely to inform all aspects of design in the foreseeable future. Green design is clearly more than the notional adaptation of familiar products: if the phrase is to be anything but empty rhetoric it will touch all aspects of raw materials, transport, pollution and manufacturing processes. It is an emotive subject and one open to many different interpretations, but there is some reason to suppose that even quite theatrically-designed green products may be necessary to galvanize public awareness of the more complex issues. Green detergents on the supermarket shelves may simultaneously preach environmental concern whilst actually stimulating consumerism, but their value in raising public awareness is significant. A similar process in product design may pull off the same trick without the stigma of continuing consumption – bins which separate garbage for recycling need only be bought once, but their continuing presence has continuing benefits. If this sort of scenario is the likely one, product designers will need to establish a whole new visual language to address this area.

Whether the product demands of social groups with special needs will be addressed with such urgency is less certain. There are already encouraging signs among more sensitive product designers that it is the similarities between special needs products and 'ordinary' ones which should be stressed, and not the differences. Style here has one of its most important roles to play. Meanwhile well-researched product design for children remains something that has curiously failed to excite manufacturers and designers outside Scandinavia. Again there is some evidence of growing interest in Great Britain and the United States, and it looks likely that this too may prove interesting territory for deploying the sort of style that relates to the consumer's actual experience and requirements.

All of which returns us to basic issues. In this discussion of style in product design I have repeatedly rejected some of the notions which are popularly associated with 'stylistic' exercises: product design as art; product design with pretentious aspirations towards being 'classic', when being good is all that should properly concern it; product design which arrogantly ignores functional flaws in the pursuit of style;

product design as merely a fashion accessory. Even so, in choosing what I consider to be good examples of style in product design, I have been very conscious that every inclusion means a dozen exclusions that might equally deserve mention. It is also inevitable that some subjectivity will have crept into value judgements, but this does not make those judgements any less useful and instructive, for subjectivity governs any consumer's choice of any product.

Sometimes identified as a kind of superfluous visual posturing, style can certainly be that. But used skilfully it can also be a positive force, not only in commercial terms but also in focusing opinions and changing attitudes. There is something of a cultural divide between a 1957 Chevrolet and a politically correct recycling bin, but style spans them both, proving that like everything else, it is in the end something to be used with wisdom or otherwise. The previous chapter gives a taste of how flexible, persuasive and positive style can be. How to stimulate, enable and achieve it remains one of the more exciting challenges for today's entrepreneurs, manufacturers and product designers.